Joseph Ato Forson

Rural Water Supply

A cost effective Analysis of selected water projects in Ghana

Document Nr. V207596

Joseph Ato Forson

Rural Water Supply

A cost effective Analysis of selected water projects in Ghana

GRIN Verlag

Die Deutsche Bibliothek verzeichnet diese Publikation in der Deutschen Nationalbibliografie; detaillierte bibliografische Daten sind im Internet über http://dnb.d-nb.de/ abrufbar.

1. Auflage 2012

http://www.grin.com
Druck und Bindung: Books on Demand GmbH, Norderstedt Germany
ISBN 978-3-656-35059-0

POLICY PROPOSAL:

RURAL WATER SUPPLY: A COST EFFECTIVE ANALYSIS OF SELECTED WATER PROJECTS IN GHANA

By:

Joseph Ato Forson

2012

Policy Issue: Lack of potable drinking water in rural Ghana

Domain of Policy: Health Policy

Category of Public Policy: Public Good

Table of Contents

Executive Summary

The axiom "*water is life and life is water*" underscores the importance of water to the everyday needs of all living things including man. The global perspective on access to safe drinking water for both domestic and agriculture needs has for some time now been a major challenge. The WHO estimates that nearly 3.4 million people die annually as a result of water and sanitation related diseases and about 99 percent of this number is from developing countries. About 780 million people lack access to potable drinking water that is one in every nine people. Women spend almost 200 million hours daily collecting water for domestic chores. These findings are jaw-dropping. Ghana, as a developing country with an estimated population of 23 million is faced with these same challenges. Incidence of water related diseases have been prevalent in most rural communities in Ghana. Background check shows that Ghana's problem in rural water supply have come as a result of low investments couple with high capital demands in carrying out annual rehabilitation works on existing facilities.

The objective of this policy proposal is to address the challenges of water supply in rural Ghana. The aim is to increase rural water access as captured in the coverage rate from 52.16% to 76% in rural communities. To achieve this, two policy alternatives (rain harvesting system and desalination plant) have been proposed alongside existing government intervention (status quo). The cost associated with each intervention has been estimated and projected to cover the duration of the proposed policy periods. To have all three interventions implemented concurrently for the duration of this policy will cost government an estimated US$66.4 million. On the other hand, making a choice among implementing any of these alternatives; borehole, rain harvesting and desalination plants will cost US$19,246,500, US$ 12,960,000 and US$22,050,000 respectively from 2012 to 2015. However, it is a policy in Ghana that rural communities should be able to provide 5% to 15% of the total project cost as mobilization cost to benefit from such facilities delivery. This component is an important factor in rural water supply. It is therefore important to look for a cost effective alternative in which the cost per head could easily be afforded. This policy factored in this aspect through the policy analysis type adopted. A cost effectiveness approach was carried on all three interventions to determine which is more expedient in terms of cost. The cost effectiveness ratio was the main criteria used to arrive at the preferred alternative. An estimate of the total annual cost was therefore needed to help compute the CER. It became necessary to compute the cost recovery factor (crf) taking into account a discount rate and the total lifespan of each policy alternative divided by the target population. The analyses showed that the per capita cost of Borehole project is between USD$0.82 – 1.38 for the next four years. Rain water harvesting system is expected to record a per capita cost ranging between USD$0.73-1.07 while that of desalination project is expected between USD$1.05-1.13. The WASHCost project funded by Bill and Malinda Gates Foundation to collect, collate and analyze cost data for water and sanitation services estimates per capita cost of small town water systems in Ghana to be between USD$12-22 and USD$4. The three alternatives have a per capita cost falling below these standards of estimation. Based on the total annual cost and the changed in improved water consumed after intervention, cost effectiveness ratio was estimated. The Cost Effectiveness Ratios (CER) for borehole project ranges from $22.03 to $30.83. Rain Harvesting System recorded a CER of $21.63 to $26.58 whiles Desalination System ranges $28.08 to $31.20. The results, tentatively reveals that Rain Harvesting System have a lower cost effectiveness ratio (CER), followed by Borehole project and Desalination plant. Therefore, rain harvesting system or projects in Ghana are more cost effective than borehole and desalination projects. However, to

make an informed decision as to which of these interventions is suitable in both long and short runs, further analyses needs to be done. Under the sensitivity analyses, the one-way varies only the discount rate while the multi-way varies both the discount rate and the lifespan of the interventions. Both analyses presents the worst case scenario reflected in the decrease in parameters and the best case scenario showing an increase in parameters. Despite the fact that the variation of the discount rate in the one-way analysis shows some changes in the CER of the three interventions, these changes are not significant enough to deviate from earlier results. In both one-way and multi-way analyses, there was a close up between borehole and rain harvesting as both were seen to have had low CER. However, in terms of consistency, rain harvesting system recorded the lowest CER throughout the analyses. A look at the NPV estimates also buttresses earlier results as the present value of rain harvesting system was least, implying given the same outcome for the three alternatives, it is economically and financially viable to invest in it. The results of the sensitivity tests/analyses reaffirms the stability of best and worst case scenarios. This policy document based on these findings strongly recommends the adoption of rain harvesting as an alternative through the Ministry of water Resource works and Housing to augment existing government's interventions but cautions that in the wake of available resources with much reference to the oil proceeds, efforts should be made to pursue desalination at the coastal parts of Ghana.

Acknowledgement

I am very grateful to Dr. Nattha Vinijinaiyapak the co-lecturer for Policy Studies (DA 802), at the National Institute of Development Administration (NIDA). This policy proposal would not have been possible without the well taught lectures and materials provided by Prof. Nattcha. The same appreciation goes to Prof.Sombat, a co-lecturer and president of NIDA for his initial introduction of the public policymaking process and the intricacies involved. This proposal would not have also been possible without the contribution of my colleagues in class. To the Ministry of water resources, works and housing and its allied agencies in the water sector in Ghana, I say thank you for the information provided. I also wish to acknowledge the endless efforts of the PhD administrative staffs of GSPA for their administrative contributions to making this course a success.

SECTION ONE

1.0 BACKGROUND

The global need for water cannot be over emphasized. The WHO and its allied agencies have been working feverishly with head of states of developing countries and other private institutions in ensuring majority of the world populace have access to good drinking water. The Millennium Development Goals (MDGs) highlights this need. According to the WHO, lack of safe drinking water has become a major concern for third world countries. Available figures indicate that more than 3.4 million people die each year from water, sanitation, and hygiene-related causes. Nearly all deaths, 99 percent, occur in the developing world (WHO 2009). It is also estimated that 780 million people lack access to an improved water source; approximately one in every nine people. Water and sanitation crisis claims more lives through disease than any war claims through gun. The same report from WHO has stated that, women spend more than 200 million hours a day collecting water (ibid). These figures are alarming. Ghana has a land size of 238,537 sq km (92,100 sq miles) with an estimated population of 23 million. Out of this 23 million, about 60% are resident in the rural communities (i.e. approximately 15 million people). The first public water supply system in Ghana, then Gold Coast, was established in Accra just before World War I. Extensions were made exclusively to other urban areas among them the colonial capital of Cape Coast, Winneba and Kumasi in the1920s. During this period, the water supply systems were managed by the Hydraulic Division of Public Works Department. With time the responsibilities of the Hydraulic Division were widened to include the planning and development of water supply systems in other parts of the country. In 1948, the Department of Rural Water Development was established to engage in the development and management of rural water supply through the drilling of bore holes and construction of wells for rural communities. Successive governments have provided Medium-Term National Development Policy Frameworks to guide the preparation and implementation of Sector and District Development Plans aimed at reducing poverty and improving the social well being of the people. The Ghana Vision 2020, Ghana Poverty Reduction Strategy (GPRS I) and the Growth and Poverty Reduction Strategy (GPRS II) are the latest of such national development policy frameworks. Throughout these policy documents, efforts have been made to mitigate the social canker of poverty through a countless number of interventions. Ghana's poverty situation has been defined from a multi dimensional perspectives ranging from security, respect, lack of social amenities, prevalence of diseases among others (GTz poverty profiling in Ghana, 2003). Ghana's poverty situation is quite pronounced in rural areas of the country where about 60% of the country's population resides and where government's interventions lack.

1.1. Present state of rural water supplies in Ghana

Gyau and Dapaah (1999) contend that about 50.5% of the rural population depends on surface waters such as streams, rivers, lakes, ponds, dams and dugouts. These sources are usually heavily polluted and are the main causes of water borne diseases so common in the rural communities. Based on the same estimates, only about 0.05% of the rural population depends on rainwater harvesting due to the unfavourable annual rainfall pattern in many parts of the country. About 40.7% depend on boreholes and wells whilst about 0.7% rely on springs for their water supply needs. Though a number of researches have been able to identify the underlying causes of most

water related diseases in Ghana, not much has been achieved in terms of the policy interventions at play. Investment in water facilities has been lower than expected. In 1984, it was estimated that rural water coverage was 55% (Gyau and Dapaah, 1999). This figure has however dwindled over the past 20 years affirming low investment giving the rising population in rural communities. Currently, the estimated rural coverage to access to potable water stands at 52.86% (CWSA Medium Term plan, 2008-2012). The Medium Term plan has also shown that the National Community Water and Sanitation Agency (NCWSA) responsible for providing rural and national water have consistently failed to meet projected targets. A look at the declining trend shows that, from an estimated 27% in 1990, rural water coverage increased to 30 % in 1999. Coverage expanded to 46.3% 2003 and 51.1% and 51.9% in 2004 and 2005 respectively. Notwithstanding this, the regional coverage has shown lot of variation across board. This trend is shown below (Strategic Investment Plan 2008-2015 & The Medium Term Plan 2008- 2012).

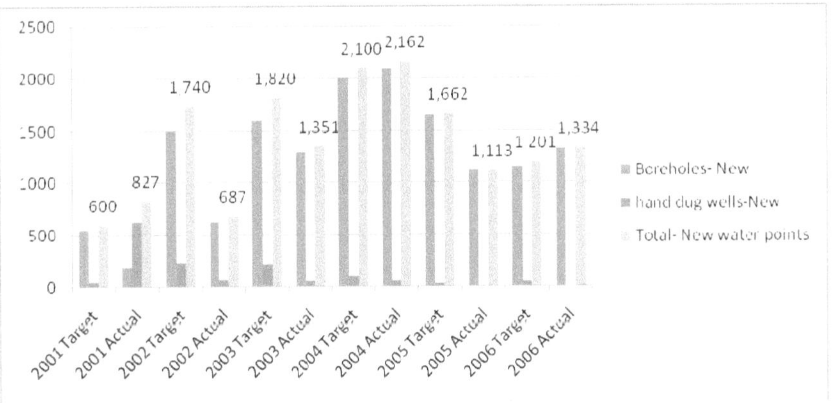

Figure 1: Facilities and service: Targets and Delivery: 2001 to 2006.
Source: Author's own construct, Data from SIP (2001-2006) & MTP (2001- 2006).

The pictorial exhibits shown above reveal that there has been a fluctuating trend with respect to facilities targets as against the actual. For instance, in 2001, the target for the construction of boreholes stood at 550. The actual provision made was 198 representing 36% with a shortfall of 64%. The period between 2001 and 2006 actuals saw an increase from 198 to 1,325 representing 569% increment. This represents a significant jump in policy interventions. However, this has not been able to meet the demand as there are countless communities without access to safe drinking water.

1.1.1 Number of New communities/small towns pipes

According to the 2008 SIP & MTP report, targets with respect to provision of small communities/ Town pipes were exceeded. For instance, in 2001, the target was to provide small

communities with 10 new pipes but the actual number of pipes provided was 92. Small towns' pipes also experienced the same scenario with actual exceeding targeted projection. Figure 2 below shows the trend:

Figure 2: small communities/towns pipes: Targets and Delivery: 2001 to 2006.
Source: Author's own construct, Data from SIP (2001-2006) & MTP (2001- 2006).

1.1.2 Rehabilitation of water facilities

The main challenge to the provision of safe drinking water in Ghana is the issue of maintenance or rehabilitating constructed water facilities. The maintenance culture in Ghana is far below expectation and this is depicted in the graph below.

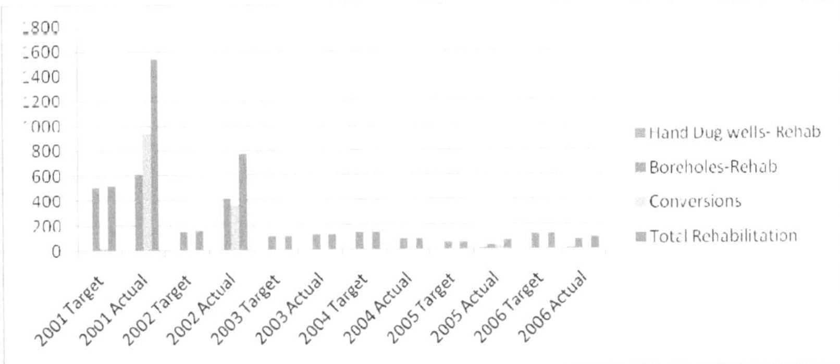

Figure 3: Rehabilitation of water facilities: Targets and Delivery: 2001 to 2006.
Source: Author's own construct, Data from SIP (2001-2006) & MTP (2001- 2006).

From fig.3, though it could be seen from the given data that actual rehabilitation has always exceeded targets, the ensuing years showed nothing to indicate if rehabilitation had taken place. In 2005, targeted rehabilitation work was 49, but only 31 of existing boreholes facilities were

actually rehabilitated. Rehabilitation works on hand dug wells were non-existence. This undoubted contributed to the continuous fall in the coverage level of rural/communities water.

1.1.3 Regional Water coverage at end-2006

The national coverage as at 2006 showed a varying degree across regional borders within Ghana. The upper west region was leading in terms of coverage, recording 67.18 %. The Ashanti, the Brong Ahafo, Greater Accra, Northern, upper east and the Volta regions recorded an average of 52% in coverage with the remaining 4 regions falling below 41.53%. See table 1 below;

Table 1: Regional Water Coverage

National Coverage							
Region	**Communities**	**Relevant Population**	**Boreholes**	**HDW**	**Pipes System**	**Population served**	**Coverage (%)**
Ashanti	2,428	2,365,244	3,483	830	73	1,491,619	**63.06**
Brong Ahafo	2,639	1,750,114	2,250	503	18	909,993	52.00
Central	3,091	1,497,292	1,053	479	25	694,189	46.36
Eastern	3,211	1,796,739	2,078	1,078	18	846,645	47.12
Gt. Accra	848	592,489	212	65	7	846,645	50.96
Northern	3,848	1,857,022	3,197	516	21	1,079,392	58.12
Upper East	1,912	1,006,078	1,633	434	6	515,855	51.27
Upper West	929	650,464	1,534	77	9	436,991	67.18
Volta	2,722	1,430,999	1,756	52	85	729,721	50.99
Western	1,739	1,440,399	1,037	418	26	598,155	41.53
Total	**23,367**	**14,386,840**	**18,233**	**4,452**	**288**	**7,604,478**	**52.86**

Source: SIP (2001-2006) & MTP (2001- 2006).

The rural water coverage mirrored/mimicked the trend in the national coverage. In terms of rural coverage, the upper west dominates with a recorded coverage of 73.39% followed by the Ashanti region which also recorded 64.26%. Most of the other regions recorded an average of 50%. The trend is further illustrated in the diagram below;

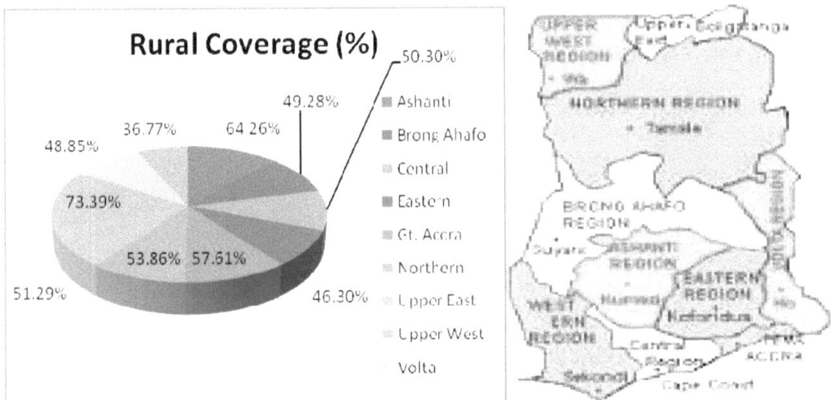

Figure 4: Rural water coverage and Administrative Regions of Ghana, 2006.
Source: Author's own construct, Data from SIP (2001-2006) & MTP (2001- 2006).

Rural water coverage as shown is far below the sustainable level needed to curb the high incidence of water related diseases. Available figures from the Ministry of Health shows that water borne diseases are among the top reported cases in Ghana. Malaria, Guinea Worm infestation, typhoid fever, diarrhea and hepatitis are among the prevalent cases (Ministry of Health, 2010). A look at the guinea worm cases across the regional borders shows that government and its development partners have to intensify interventions in eradicating guinea worm in Ghana.

Table 2: Guinea Worm cases in Ghana from 2001-2006

Region	2006	2005	2004	2003	2002	2001
Ashanti	53	59	85	45	39	50
Brong Ahafo	204	289	336	492	779	860
Central	0	0	0	0	9	11
Eastern	8	17	1.604	1.511	305	595
Greater Accra	1	3	3	3	3	6
Northern	3679	2973	4.979	5.999	4.207	2.929
Upper East	6	7	222	152	128	186
Upper West	90	322	15	23	29	13
Volta	86	286	27	37	45	83
Western	2	2	4	28	1	5
National	**4,129**	**3958**	**7,275**	**8,290**	**5,545**	**4,738**

Source: Guinea Worm Eradication Programme

Figure 5: Guinea Worm Eradication program

The Northern region of Ghana has the highest guinea worm infestation cases, followed by the Eastern region. Government's interventions must focus on these regions where the prevalence rates in water related disease is high. Ghana reported 5,497 cases of dracunculiasis in 741 villages in 2002, or 11% of the global total for that year. 95% of the cases occurred in only 15 of the country's 110 districts. Five of Ghana's 10 regions (Ashanti, Central, Greater Accra, Upper East and Western) reported only imported cases in 2002, all of them reportedly contained. Ghana reported 15% more cases in January 2003 (859) than in January 2002 (744). The epicenter of Ghana's remaining endemic area is in the eastern part of the Northern Region, encompassing land that is very fertile—including the three top yam-producing districts in the country—and draws migratory seasonal farmers from many other areas of Ghana.

Table 3: coverage by interventions in Ghana-2001-2002

Interventions	2001	2002
Endemic village with 100% filters	85	95
Endemic village using Abate	20	26
Endemic village with 1+ safe water	34	44
Percentage of cases contained	68	66
Number of cases reported	5,344	5,497
Number of villages reporting 1+ case	784	741

Source: Public Health Service center for Disease Control, GHS (Ministry of Health, 2006)

1.2 Climate

The climate of Ghana which is of primary importance in understanding the spatial and temporal distribution of surface waters is influenced by three air masses namely, the South-West Monsoon, the North-East Trade Winds (Tropical Continental Air Mass) and the Equatorial Easterly. The warm but moist South West Monsoon which originate from the Atlantic Ocean and the warm, dry and dusty Tropical Continental Air Mass (Harmattan) from the Sahara Desert approach the tropics from opposite sides of the equator and flow towards each other into a low pressure belt known as the Inter Tropical Convergence Zone (ITCZ). The slow and irregular north-south oscillations of the ITCZ gives rise to the regime of wet and dry seasons. The wet season in the southern sections of Ghana is characterized by two main rainfall regimes, i.e., double maxima whilst the northern sections experience single rainfall regime in a year. The extreme south-western portion of Ghana is the wettest part of the country which receives more than 2000m of rainfall a year. Rainfall which mainly recharges the aquifers generally decreases towards the north and south-eastern sections of the country. The driest part of the country is found in the south-east coastland plains where the mean annual rainfall is about 800mm. Mean monthly temperature over the country never falls below about 25°C while open water (pan) evaporation is generally high and ranges from about 1200mm per year in the south-west to more than 2600mm in the north. Relative humidity's are high on the coast and are generally between 95% and 100% during the night and early morning. These can reach low values of between 20% and 30% or less in the north when the area comes under the influence of the dry Tropical Continental Air Mass (Gyau and Dapaah 1999).

1.3 Geology

The country is underlain partly by what is known as the Basement Complex which comprises a wide variety of Precambrian igneous and metamorphic rocks. These crystalline rocks cover about 54 percent of the country. They can be further divided into subregions on the-basis of geology and groundwater conditions. These consist mainly of gneiss, phyllites, schists, migmatites, granite-gneiss and quartzites. About 45 percent of the country is underlain by Palaeozoic consolidated sedimentary rocks locally referred to as the Voltaian Formation and consist mainly of sandstones, shale, arkose, mudstone, sandy and pebbly beds and limestones. Both the Basement Complex and the Voltaian formation have little or no primary porosity, hence groundwater occurrence is associated with the development of secondary porosity resulting from jointing, shearing, fracturing and weathering. This has consequently given rise to two main types of aquifers which are the weathered zone and the fractured zone aquifers. The remaining 1% of the rock formations are associated with aquifer formations and are made up of Cenozoic and Mesozoic sediments which consist of unconsolidated alluvial sediments, beach sand, red continental deposits of mainly alternating limonitic sand, sandy clay gravels, marine shale, limestone and glauconitic sandstone[1].

1.4 Snapshot of the National Community Water and Sanitation Agency in Ghana

The community water and sanitation agency is one of the agencies under the Ministry of Water Resources, Works and Housing (MWRWH). The National Community Water and Sanitation Agency was launched in 1994 to oversee to community and rural water provision. To ensure the implementation the community water and sanitation agency was subsequently established by an Act of parliament (ACT 564 of 1998) to facilitate the provision of safe water and related sanitation services to rural communities and to provide for connected purposes. CWSA works with District Assemblies and communities to achieve its objectives. Its primary vision is to be a highly effective and efficient professional organization working harmoniously with stakeholders to deliver sustainable water and related sanitations to all rural communities and small towns. The agency has links with other organizations within the water sector as provided below;

- Water Resources Commission,
- Water Research institute,
- Ghana Water company Limited,
- the Ministries of Health, Education, Finance and Economic Planning,
- Local Government,
- Rural Development and Environment,
- Ghana Standards Board,
- the Environmental Protection Agency,
- Support Agencies,
- Partner Organizations (Consultants, Contractors),
- NGOs,
- The District Assemblies and the beneficiary Communities in rural and small towns.

[1] See Gyau and Dapaah (1999).

SECTION TWO

2.0 LITERATURE REVIEW

The purpose of this literature review is to appraise well-known interventions for dealing with the provision of potable water for rural and small towns especially in the vein of increasing access and eradicating water borne diseases. In the literature review, we will try to estimate the disease burden related to water, sanitation and hygiene and then also assert the benefits explored with respect to reliance on groundwater (Hand Dug well/ Boreholes), Rain harvesting (water tanks) and desalination. It should be mentioned that interventions such as construction of dams, construction of boreholes or wells, trucking of water and pipe system are not explored in this literature as they are already embodied in existing Government's interventions in Ghana.

The global perspective of water shortage, contamination and related diseases as well as sanitation is alarming according to the WHO report. It is estimated that close to 2.5 billion people in developing countries lack improved sanitation facilities, and nearly one billion people do not have access to safe drinking water. These unsanitary environments allow diarrhea-causing pathogens to spread more easily (WHO 2009). The report also asserts that Africa and South Asia account for one half the cases of childhood diarrhea. It is estimated that 1.4 million preventable cases are reported dead each year due to cases of diarrhea. Treating diarrhea cases and other water related disease have been thought to be more expensive.

According to Fewtrell et al (2007), there are 2 billion cases of Intestinal infection globally that could be prevented .Transmission of intestinal nematode infections, such as ascariasis, trichuriasis and hookworm, occurs through soil contaminated with faeces. It is entirely preventable by adequate sanitation, and intervention outcomes are reinforced by good hygiene. In our estimates, the burden caused by intestinal nematode infections is, therefore, entirely attributable to inadequate sanitation facilities and related lack of hygiene. Lymphatic filariasis has seriously incapacitated 25 million people globally. In Asia and the Americas, lymphatic filariasis is transmitted by mosquito vectors breeding in water polluted by organic material, and its distribution is therefore linked to urban and periurban areas. In Africa, where *Anopheles* mosquitoes are the main vector, its distribution coincides in part with that of malaria and may be linked to irrigation development. Lymphatic filariasis also occurs in some of the Pacific island states. Globally, 66% of the disease is attributable to unsafe water, inadequate sanitation or insufficient hygiene. Tahoma impairs nearly 5 million people annually that could have been prevented. Trachoma is a contagious eye disease that can result in blindness. It is transmitted primarily as a result of inadequate hygiene, and transmission can be reduced by facial cleanliness, access to safe water, adequate sanitation facilities and fly control. In practice, the burden caused by blinding trachoma can be almost fully attributed to unsafe water, inadequate sanitation or insufficient hygiene. According to the WHO, there are 200 million people infested with Schistosomiasis. Schistosomiasis is caused by contact with water bodies contaminated with the excreta of infected people and is therefore fully attributable to unsafe water, inadequate sanitation or insufficient hygiene. Its distribution is linked to the distribution of the aquatic snails that are the intermediate hosts of the parasitic trematode flatworms. Along with snail control, the provision of safe water and sanitary facilities would limit infective water contact and contamination of the environment and greatly reduce the incidence of this disease (*ibid*). See figure 7 below;

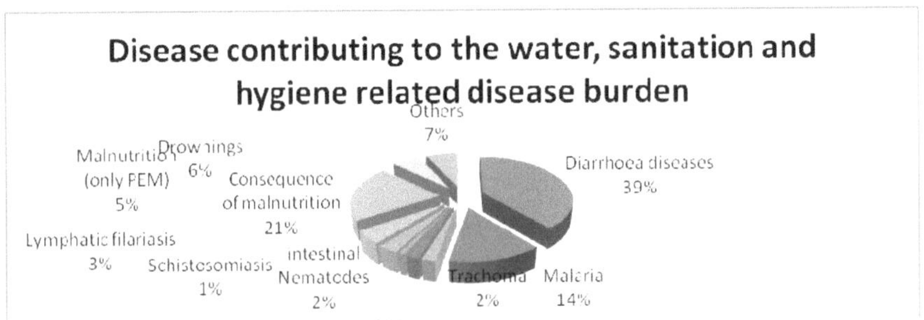

Figure 6: Water related Diseases globally
Source: Author's own construct based on data from WHO (2009)

The figures being mentioned are alarming and calls for efforts to be devised to reduce these occurrences.

2.1 The Search for Interventions

The global perspective and the urgency to address provision of safe drinking water, sanitation and hygience as a developmental and public health issue have been established. The search for interventions to address the seemingly insurmountable problem predates as far back as 30years ago. With the global climatic conditions changing coupled with its attendant's effects on water resources calls for measures through interventions to avert the challenges associated with water related diseases. Keiser et al (2005) in an attempt to find an alternative measure to reduce malaria reviewed literature of 40 case studies that emphasized environmental management interventions. Of these 40 case studies environmental modification (measures aiming to create a permanent or long-lasting effect on land, water, or vegetation to reduce vector habitats—e.g., the installation and maintenance of drains) was the central feature in 27 studies, environmental manipulation (methods creating temporary unfavourable conditions for the vector— e.g., water or vegetation management) in four, and nine quantified the effect of modifications of human habitation. They conclude that malaria control programmes that emphasize environmental management are highly effective in reducing morbidity and mortality.

2.2 Alternative sources of water

2.2.1 Rainwater Harvesting

Besides the intervention discussed by Keiser et al (2005), there are other interventions or policy alternatives that policy makers can adopt in solving the social problem associated with water, interventions like rainwater harvesting has been successfully explored by millions of inhabitants and in facts happen to be the oldest source dating back to 4000 years ago. Faber Maunsell (2004) in a feasibility study to explore rainwater harvesting and recycling of what it termed as greywater in the Birmingham Eastside emphasized that this intervention is highly sustainable especially in the long run. They also assert that in terms of cost, rainwater is much more efficient as rainwater doesn't require any form of treatment. In past centuries, river water (and often ground water)

was contaminated by upstream use of the river for washing and sewage disposal, leading to outbreaks of cholera and dysentery in the UK as late as the 19th century(ibid). Rainwater storage was seen to offer a cleaner and less risky source of water, although the reasons for the diseases were not linked to contaminated water until relatively recently. Perhaps because of this distant memory, rainwater is still perceived by the majority of people to be a clean, fresh source of water, and therefore it is less subject to objection than greywater systems. There are examples in the UK of homes that use rainwater for all purposes including drinking. However these are few and the majority of rainwater systems can be divided into three types: Irrigation systems, systems supplying WCs and washing machines, and process systems. Rainwater technology varies in the UK in accordance with the storage facilities such as storage system, including sub-surface and roof storage, and storage in balancing ponds.

Irrigation systems - Irrigation systems at the domestic level are familiar to all, being a simple water butt with a downpipe connector to collect surface water running from the roof of the property. The water can be removed either by connecting a hose to the butt, or by using a watering can. These simple systems require little maintenance and are very useful for irrigation or for washing cars and other such uses. They are limited by their relatively small storage volume which means that there is rarely sufficient water for use throughout the whole dry summer period.

WC and washing machine systems - These systems are more sophisticated as they require dual plumbing to be installed in the building, supplying the WC and washing machine, whilst other water uses are supplied by the standard mains water. In order to provide sufficient storage of Rainwater to make the system viable, a large storage tank is generally installed. Storage at high level is possible, but with the majority of systems available, an underground storage tank is used, which feeds a small header tank at high level. The diagram below shows this type of system applied to a house.

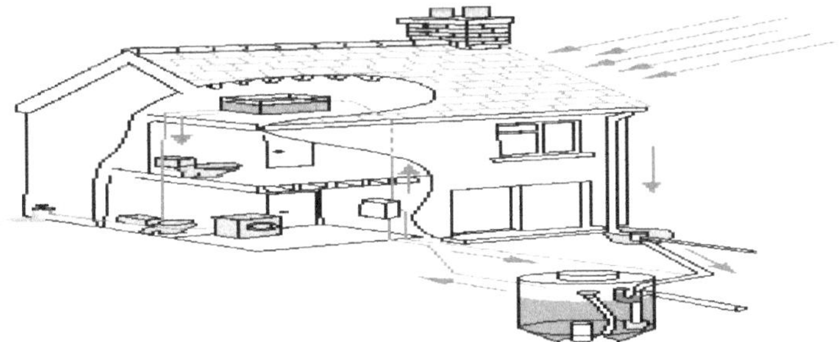

Figure 7: *WC and washing machine systems*

Source: Faber Maunsell (2004)

From the on-going description, it can be established that rainwater harvesting in the UK has been very fruitful except that when this technology is introduce in any of the developing countries, the cost involved won't make it sustainable especially in the short run. However, other cost effective

technologies have been successfully developed and used in other developing countries like Pakistan and India.

Muhammad Akram Kahlown (2001) explores how rainwater harvesting has been successfully researched in the Cholistan desert under the Pakistan Council of Research in Water Resources (PCRWR). Under this research programme, the PCRWR has developed 92 rainwater harvesting systems on pilot scale in Cholistan desert. Each system consists of storage reservoir, energy dissipater, silting basin, lined channel, and network of ditches in the watershed. The storage pond is designed to collect about 15000 m3 of water with a depth of 6 m. Polyethylene sheet (0.127 mm) on bed and plastering of mortar (3.81 cm) on sides of the pond was provided to minimize seepage losses. All these pilot activities to harvest rain have brought revolution in the socio-economic uplift of the community. These activities have also saved million of rupees during the recent drought. Large scales adoption of all these interventions would ultimately help improve the socio-economic conditions of the residents of hyper arid area of the country. Though this programme is a pilot one, the mere fact it has recorded an unprecedented success in a seemingly desert area could be a panacea to other developing countries like Ghana where the climate is favorable in terms of rainfall pattern. Also, the construction materials used by the Pakistani pilot project were mainly local materials which abound in most developing countries.

Similar project have been successfully explored in India under the "Artificial Recharge" and "rainwater harvesting" initiatives. Groundwater in India have been thought to be over exploited since the annual ground water extraction exceeds the annual replenishable resource and significant decline in long term ground water levels has been observed either in pre- monsoon or post-monsoon or both. The continuous over exploitation has called for other alternatives such as rain harvesting to be considered (Central Ground Water Board, 2001:2). This has been well executed and has benefited millions of inhabitants in the region (ibid).

It should be mentioned that the components of rainwater harvesting differs across countries. The Shumacher Center for Technology and Development (2006: 4) asserts that rainwater is the easiest and cost effective way of collecting water but is less relied on due to lack of information. The report elaborates on the cost effective materials used in building water storage facilities in Sri Lanka and Tanzania. It also gives a detail account of the pros and cons of using tank and cistern. There are varying methods of sizing a tank. The report makes reference to three distinct methods: the *demand side approach* which is a very simple method and usually calculates the largest storage requirement based on the consumption rates and occupancy of the building. This approach takes into account the consumption per capita per day, number of people per household, and longest average dry period. The second method is *the supply side approach*. The second approach is appropriate to serve larger community or big hospital etc. The third approach is the *computer models* with software programmes such as SimTanka that does the calculation of the size of the tank more accurately (ibid: 11). Gould and Nissen-Peterson (1999) emphasized on rain water harvesting as an alternative for solving the water challenges and categorized rainwater harvesting according to the type of catchment surface used and the scale of activity and went further to have it presented in the diagram below;

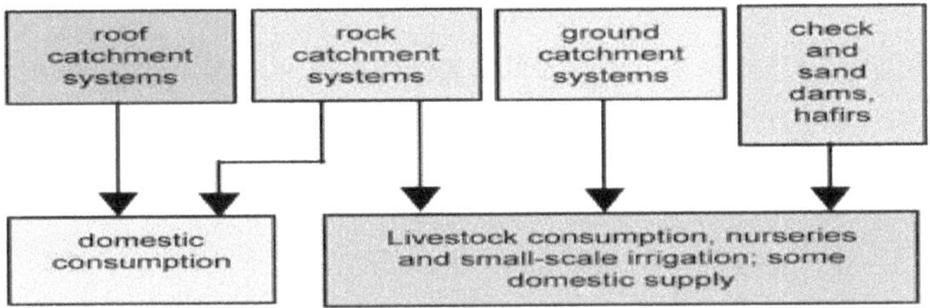

Figure 8: Small-scale rainwater harvesting systems and uses (adapted from Gould and Nissen-Petersen, 1999).
Source: cited from Norma Khoury-Nolde (2006).

Rooftop rainwater harvesting at the household level is most commonly used for domestic purposes. It is popular as a household option as the water source is close to people and thus requires a minimum of energy to collect it. An added advantage is that users own maintain and control their system without the need to rely on other community members (Norma Khoury-Nolde, 2004: 2).

2.2.2 Desalination

Another alternative of water supply is desalination, which is the removal of salt content in existing water source. Desalination is one of the most expensive means of getting water according to a case study report of Ashlynn Stillwell (2008:9). Not only being expensive, the process is time consuming as more energy is required to overcome friction. Desalination requires 9,780 to 16,500 kilowatt-hours per million gallons (kWh/Mgal). Using an average value of 13,140 kWh/Mgal for treatment of 20 mgd, the energy requirements for desalination total 262,800 kWh/day. In Limassol-Cyprus, this approach has been used as a cost effective alternative to supplying water. The cost effective plant supplies nearly 22,000m^3 of water on daily basis to the city of Limassol (Nirosoft treating water worldwide, 2008). Nirosoft designed, manufactured, and installed a containerized reverse osmosis desalination system with a capacity of 4,250 m3/day. The reliable system operates 24 hours per day, producing high quality water all year round (ibid) see figure 10 below;

Figure 9: Containerized seawater desalination in Cyprus/Cape Verde.
Source: (Nirosoft treating water worldwide, 2008)

Nitrosoft has designed similar plants in Kenya for farmers' irrigation purposes using a cost effective approach (Nirosoft treating water worldwide, 2008). In Minjur, India, similar technology has been developed at a relatively cheaper cost. The technology has been thought to be environmentally friendly and an alternative to reliance on rainy water giving the changing trend in rainfall pattern and shortages brought about by drying water sources (Energy Recovery Inc., 2011).

2.2.3 Borehole/ Hand Dug Well

Increasing access to groundwater is one of the top most priorities of most developing countries especially in the sub-Saharan Africa. One key note is to reduce the cost of conventional drilling and borehole construction. The usual perception held in connection to borehole construction is that the approach is very expensive. It's been estimated that Africa needs about one million boreholes to reach full coverage[2]. The studies continuous that if a 10 percent reduction could be achieved on the average US$10,000-15,000 cost of a borehole in much of Africa, this would result in savings of more than US$1 billion. This potentially adds another 30 million people with access to safe water. The Ethiopian case study unraveled an important feature in borehole drilling in which it states that drilling should not be too deep in a bid to reduce cost associated with drilling deep.[3] However, one of disadvantages raised has to do with the fact that, every 100 boreholes drilled, 70 will be successful and 30 will be dry. There is therefore the need to work within a back-up plan ascribed by Carter et al (2006) who calls for allowance in the estimate to make up for the dry boreholes.

Figure 10: Borehole and a Drilling machine in Ethiopia
Source: WSP (2006)

The costing of a complete borehole involves a number of items that usually makes it expensive. Data on the groundwater has to be collected; estimate the depth of water on an existing water table among others. In Ghana, Sudan and most developing countries, this alternative is the most common intervention used to increase accessibility. However, the high cost of maintenance and the associated perennial changes in rainfall pattern leads to dry-up of most wells thereby increasing the challenges with meeting goal of accessibility.

---

[2] Cited from Water and Sanitation Program-Ethiopia, (2006:2).
[3] ibid

2.3 Wrap-up of literature reviewed

The literature review has underscored the efficacy of the three well-known policy interventions (Borehole, Rainwater Harvesting and Desalination) in dealing with the challenges of rural water supply. These will be simulated within the Ghanaian context to ascertain their cost-effectiveness in dealing with the persistent and seemingly insurmountable challenges of inadequate water supply and its associated socio-economic effect.

SECTION THREE

3.0 Policy Goal

The goal of this policy proposal is to harmonize government effort in enhancing access to safe drinking water in line with Act 564 of 1998, which saw to the establishment of Community Water and Sanitation Agency (CWSA) to be responsible for the management of rural water supply systems, hygiene education and provision of sanitary facilities. After the establishment of CWSA, 120 water supply systems serving small towns and rural communities were transferred to the District Assemblies and Communities to manage under the community-ownership and management scheme[4].

3.1 The research questions for this policy proposal

The key question of this research is; what are the internationally accepted cost-effective interventions for dealing with the provision of safe drinking water and sanitation? In relation to this question, groundwater (boreholes/wells), rain harvesting and desalination are the known interventions. These were further explored to unearth their effectiveness in dealing with safe drinking water and its attendant's related diseases in Ghana. Among some of the questions addressed from these perspectives are;

- Will increasing access to safe drinking water through the interventions avert the huge socio-economic burden to the nation as a result of water borne diseases?
- In terms of cost which of the three policy interventions should be recommended to the sector ministry?

3.2 Objectives of the Policy

The general drive of this policy proposal is to raise the issue of rural water provision as public policy issue and to also present to policy elites the alternative solutions for dealing with the challenges based on authoritative evidence. Also, this policy proposal will seek to achieve the following specific objectives;

- To increase rural population coverage of access to safe drinking water from 52.86% to 76% by 2015.
- To reduce the incidence of water related diseases by 85% come 2015.

3.3 Policy alternatives

The identification and specification of policy alternative is one component of policy analysis. There are a number of ways policy alternatives can be developed, for example, from existing policy proposals; policies implemented in other jurisdictions; generic policy solutions and custom-designed alternatives. In this paper, the policy alternatives outlined are a mixture of these sources. This paper identifies three policy alternatives to enhance sustainable accessibility and supply of water to rural communities. The following have been thought of;

---

[4] See http://www.cwsagh.org/

	Status quo: Reliance on groundwater (Borehole/Hand dug wells)
Option 1	Provision of Groundwater (BH/HDW) ✓ *Construction of boreholes or hand dug wells for rural communities*
Option 2	Rain Harvesting System (RWS) ✓ *Distribute water tanks to rural communities*
Option 3	Desalination Plant (DP)

3.4 Scope and Methodology

Rural water supply can be verified from four differing broad areas such as human, access to potable water, reduction in water related diseases and the environment. Incorporated in these four broad areas is the construction and management of water facilities which in most documents have been labeled as hardware and software component. This proposal focuses on water facilities aspect in a bid to increase access and curb the incidence of water related diseases. Succinctly, three interventions are analyzed which are borehole/hand dug well, rain water harvesting and desalination. The scope however does not cover the software component into detail though the costing of these facilities takes account the software aspect.

In terms of data, the study obtained relevant secondary but authoritative data from the community water and sanitation agency, Ghana. Additionally, further information was obtained from the internet from notable organizations such as the World Health Organization, ISODEC to support relevant literature for a detailed analysis.

3.5 Limitations

The only limitation with this study has to do with the fact that, it relied on secondary data which in actually fact were not up to date. In furtherance, prices quoted are based on literature which might not reflect the actual prices on the ground. Literature on public policy formulation as captured in the policy process calls for stakeholders participation, therefore a policy proposal of this nature must leave room for stakeholder participation. In addition, we would have wished to include rainfall pattern in our analysis as one of the parameters in the sensitivity analysis, but we were constraint by with difficulty in getting the exact figures from the metrological service department in Ghana. We also needed a formula that could accept this parameter as a component but due to time constraint, we were unable.

3.6 How to Measure

Policy criterion provides a basis for measuring progress toward achieving a goal. The measure of the successful implementation of each policy alternative will be the averted incidence of water-borne diseases in the prevalent areas, increased water access coverage from 52.86% to 76% as per targets set from 2012 – 2015.

3.7 Type of Analysis

The study employs financial analysis type specifically cost-effective approach to appraising all the three policy alternatives in relation to the goal to be attained. Cost-effectiveness analysis (CEA) is widely used to assess the investment criteria where cost-benefit analysis (CBA) is not economically feasible to apply. CBA is generally used where there is a commercial motive whereas CEA is used where service motive is prioritized. The major benchmark for a cost effectiveness analysis is the cost effectiveness ratio which is obtained by dividing the total cost of an intervention by an effectiveness measure which is usually based on the outcome of the intervention. This effectiveness measure could be in terms of the change in average water usage as a result of the intervention, time savings or diarrhea cases averted for water interventions. For educational projects, the effectiveness measure could be change in enrollment and for health projects number of deaths averted among others. The cost effectiveness ratios of two or more interventions are usually compared to ascertain the most effectiveness option. A relative lower cost effectiveness ratio represents a better investment option all things being equal. However, it is also important to note that cost-effectiveness analysis (CEA) has its own demerits and challenges such as the complexity in monetizing all gains, whose health or live to be saved should be forgone as a result of meeting efficiency gains, variations in use of discount rate, among others. In the area of water and sanitation, CEA is increasingly becoming popular. Julio Berbel et al (2010) and Dr. Thomas Clasen (2008) have both employed the approach in the European context in relation to water disease prevention and water saving perspectives. In the case of this paper, our focus is on the total cost expended in each intervention to maximize the same gains which could be relied on as basis for policy decision makers. The cost-effectiveness ratio as the change in improved water consumed per household per year by policy intervention per unit cost of implementing the policy intervention (ISODEC, 2011).

Cost-Effectiveness = change in improved water consumed per household per year

Cost-effective ratio= total annual cost of interventions / change in improved water consumed per houschold per year.

SECTION FOUR

4. ANALYSIS OF POLICY ALTERNATIVES

This section takes a critical look at the three policy alternative types by making use of data from Ministry of Water Resources, Works and Housing, ISODEC, and Community water and sanitation Agency and the literature reviewed. It is assumed that the policy implementation will take effect from 2012 to 2015. To that effect, all data analysed are from 2012 to 2015. The cost data have all been converted to US$. We begin by looking at the target projection for facilities delivery in terms of coverage from 2012 to 2015.

Table 3: Targets of facilities delivery, 2012-2015

Type	76% facility coverage	2012	2013	2014	2015	TOTAL
WATER	Borehole	583	675	727	976	2,961
WATER	Hand dug Wells	108	46	78	96	328
WATER	Rain Harvesting System (Rural)	44	43	75	54	216
WATER	Desalination Plant (Small Town)	18	15	19	11	63
SOFTWARE	Water and Sanitation committee	1,733	2,079	2,495	2,994	9,301
SOFTWARE	District water and Sanitation Teams	16	20	23	28	87
SOFTWARE	Technical Assistance	33	39	47	56	175
SOFTWARE	Small Towns consultancy service	16	20	23	28	87

Source: Authors construct based on SIP projections, (2012-2015).

This analysis envisages increasing the coverage rate in water delivery facilities in Ghana from 52.86 % to 76% (i.e. 23.14% for the intervention period) by the end of 2015. From the projection table, it is estimated that the number of boreholes is expected to increase from 583 to 979 nationwide by 2015, bringing the total number of boreholes to 2,961. Facilities delivery on the others (i.e. Rain Harvesting Systems, Desalination Plant) is expected not to increase exponentlally as the projections indicate a fluctuating trend from the table above.

Table 4: Projected Cost USD$ of facilities required to meet targets, (2012-2015)

Type	76% facility coverage	2012	2013	2014	2015	TOTAL
WATER	Borehole	3,789,500	4,387,500	4,725,500	6,344,000	19,246,500
WATER	Hand dug Wells	324,000	138,000	234,000	288,000	984,000
WATER	Rain Harvesting Tanks(Rural)	2,640,000	2,580,000	4,500,000	3,240,000	12,960,000
WATER	Desalination Plant (Small Town)	6,300,000	5,250,000	6,650,000	3,850,000	22,050,000
	Sub-total Water	13,053,500	12,355,500	16,109,500	13,722,000	55,240,500
	Project Management	652,675	617,775	805,475	686,100	2,762,025
	Hardware sub-total	**13,679,175**	**12,973,275**	**16,914,975**	**14,408,100**	**57,975,525**
SOFTWARE	Water and Sanitation committee	1,039,800	1,247,400	1,497,000	1,796,400	5,580,600
SOFTWARE	District water and Sanitation Teams	8,000	10,000	11,500	14,000	43,500
SOFTWARE	Technical Assistance	33,000	39.000	47,000	56,000	175,000
SOFTWARE	Small Towns consultancy service	480,000	600,000	690,000	840,000	2,610,000
	Sub-total software	**1,500,000**	**1,896,400**	**2,245,500**	**2,706,400**	**8,348,300**
	TOTAL COST@76%	15,266,975	14,869,675	19,160,475	17,114,500	66,411,625

Source: Authors construct based on SIP projections, (2012-2015).

For the next four years, the total cost of attaining a 76% coverage is estimated at USD$66,411,625. Desalination system for small town is expected to cost USD$22,050,000

followed by Borehole also estimated to cost USD$19,246,500 of the total cost. The software component is expected to cost USD$8, 348,300 of the total cost.

Table 3: Targeted Rural population for interventions

Relevant population	Population served	Target Population	Targeted population for the 4 years period	Relevant Communities	Communities served	Target Communities
11,630,823	5,066,750	6, 564,073	1,564,875	23,081	12,351.80	11,015.20

Source: Authors construct based on CWSA guideline, (2010).

Of the relevant rural population of 11,630,823, about 5,066,750 have been served with access to potable water representing 52.16%. The remaining 47.84% (6,564,073) are not served; implying previous interventions have so far not caught up with them. This population is the focus of this proposal.

According to the CWSA guideline in costing a project, the following have been outlined as the main constituents of each project. The components/ingredient are further stated below in table 4;

Table 4: Cost ingredients of three interventions: Borehole/ Rain harvesting/ Desalination Plant

Borehole, Rain Harvesting and Desalination Projects
1. construction
2. community mobilization and training cost
3. personnel cost
4. operation and maintenance cost

Source: Authors construct based on CWSA guideline, (2010).

The guideline makes two assumptions[5]: that some of the cost components are for the lifespan of the intervention while others are annual cost. The construction cost and the community mobilization and training cost of the interventions are for the entire lifespan of the project, while the operation and maintenance and personnel are annual cost. It is therefore necessary to calculate the annualized construction and mobilization cost. The formula[6] below is therefore used;

Calculation of annualized Project Cost
Capital recovery factor (crf)[7] $= [r(1+r)^y]/[(1+r)^y-1]$ Where r= discount rate and y= years i.e. lifespan of intervention

r= 10% and y=15 years for small town desalination plant, Boreholes and Rain Harvesting Plant.

Therefore $= [0.1(1.1)^{15}]/ [(1.1)^{15}-1]$

$= 0.418/3.177$

$= 0.13$

[5] Average household size of most rural communities in Ghana is 8. Average water delivered before borehole intervention is negligible and assumed to be zero (0). Most communities depend on rivers, streams and other unimproved sources of water (see appendix for calculations).

[6] Adopted from CWSA Guidelines as cited from ISODEC report (2010)

[7] Construction cost * crf = annualized construction cost

Mobilization & Training cost *crf= annualized mob. & training cost

Table 6: Calculation of Total Cost of Interventions/ Alternatives in USD$

m	Borehole Project-*Status Quo*				Rain Harvesting				Desalination			
	2012	2013	2014	2015	2012	2013	2014	2015	2012	2013	2014	2015
pulation	**1,564,875**	**1,471,608**	**1,383,900**	**1,301,420**	**1,564,875**	**1,471,608**	**1,383,900**	**1,301,420**	**1,564,875**	**1,471,608**	**1,383,900**	**1,301,420**
nstruction st	3,789,500	4,387,500	4,725,500	6,344,000	2,640,000	2,580,000	4,500,000	3,240,000	6,300,000	5,250,000	6,650,000	3,850,000
n. nstruction t	492,635	570,375	614,315	824,720	343,200	335,400	585,000	421,200	819,000	682,500	864,500	500,500
mm. Mob. Tr.	1,039,800	1,247,400	1,497,000	1,796,400	1,039,800	1,247,400	1,497,000	1,796,400	1,039,800	1,247,400	1,497,000	1,796,400
nualized m.Mob. st	135,174	162,162	194,610	233,532	135,174	162,162	194,610	233,532	135,174	162,162	194,610	233,532
rsonnel	652,675	617,775	805,475	686,100	652,675	617,775	805,475	686,100	652,675	617,775	805,475	686,100
e.& Main.	33,000	39,000	47,000	56,000	33,000	39,000	47,000	56,000	33,000	39,000	47,000	56,000
tal Annual st	**1,286,484**	**1,389,312**	**1,661,400**	**1,800,352**	**1,137,049**	**1,154,337**	**1,632,085**	**1,396,832**	**1,639,849**	**1,501,437**	**1,911,585**	**1,476,132**
capita st	0.82	0.94	1.20	1.38	0.73	0.78	1.18	1.07	1.05	1.02	1.38	1.13

Source: Author's construct, 2012.

The per capita cost of the Borehole project is between USD$0.82 – 1.38 for the next four years. Rain water harvesting system is expected to record a per capita cost ranging between USD$0.73-1.07 while that of desalination project is expected between USD$1.05-1.13. The WASHCost project funded by Bill and Malinda Gates Foundation to collect and analyze cost data for water and sanitation services estimates per capita cost of small town water systems in Ghana to be between USD$12-22 and USD$4. The three alternatives have a per capita cost falling below these standards of estimation.

Table 7: Calculation of Cost Effectiveness Ratio

	Borehole Project-*Status Quo*				Rain Harvesting System				Desalination Plant			
	2012	2013	2014	2015	2012	2013	2014	2015	2012	2013	2014	2015
l ual Cost	**1,286,484**	**1,389,312**	**1,661,400**	**1,800,352**	**1,137,049**	**1,154,337**	**1,632,085**	**1,396,832**	**1,639,849**	**1,501,437**	**1,911,585**	**1,476,132**
lation	1,564,875	1,471,608	1,383,900	1,301,420	1,564,875	1,471,608	1,383,900	1,301,420	1,564,875	1,471,608	1,383,900	1,301,420
of Hsld n/HH	195,609.38	195,609.38	195,609.38	195,609.38	195,609.38	195,609.38	195,609.38	195,609.38	195,609.38	195,609.38	195,609.38	195,609.38
Water umed (in s)/ hsld before vention	0	0	0	0	29, 200	29, 200	29, 200	29, 200	29, 200	29, 200	29, 200	290, 200
.water umed (in s)/househ /year vention	58,400	58,400	58,400	58,400	81,760	81,760	81,760	81,760	81,760	81,760	81,760	81,760
nge in roved er sumed sehold year	**58,400**	**58,400**	**58,400**	**58,400**	**52,560**	**52,560**	**52,560**	**52,560**	**52,560**	**52,560**	**52,560**	**52,560**
D$/litre)	22.03	23.79	28.45	30.83	21.63	21.96	31.05	26.58	31.20	28.57	36.37	28.08
NKING	**2nd**				**1st**				**3rd**			

Source: Author's construct, 2012.

4.1 Discussion of Results

The Cost Effectiveness Ratios (CER) for borehole project ranges from $22.03 to $30.83. Rain Harvesting System recorded a CER of $21.63 to $26.58 whiles Desalination System ranges from $28.08 to $31.20.

The results according to the ranking done, tentatively reveals that Rain Harvesting System have a lower cost effectiveness ratio (CER), followed by Borehole project and Desalination plant. Therefore, rain harvesting system or projects in Ghana are more cost effective than borehole and desalination projects. However, we are unable to base our argument on this to make an informed decision as to which of these interventions is suitable in both the long and short run unless further analysis is done varying the parameters initially used and more also considering the net present values of all the three interventions cost as the basis assuming the investment is to be done today, based on limited financial resource, which should be selected to generate the same level of outcome.

4.2 Sensitivity Analysis

Under the sensitivity analysis, we test the stability of the results in table 7 and show significant or insignificant the CER changes as various parameters employed are varied to ascertain the effects of such changes on the overall results. One-way and multi-way sensitivity analyses[8] are presented in table 8 below;

Table 8: One-Way and Multi-Way Sensitivity Analysis on CEA ***(table continues on next page)***

Sensitivity On Discount Rate												
Base Case: 10%												
	Borehole-*Status Quo*				**Rain Harvesting System**				**Desalination Plant**			
	2012	2013	2014	2015	2012	2013	2014	2015	2012	2013	2014	2015
Total Annual Cost (US$)	**1,286,484**	**1,389,312**	**1,661,400**	**1,800,352**	**1,137,049**	**1,154,337**	**1,632,085**	**1,396,832**	**1,639,849**	**1,501,437**	**1,911,585**	**1,476,132**
Change in Water Consumed per household/yr	58,400	58,400	58,400	58,400	52,560	52,560	52,560	52,560	52,560	52,560	52,560	52,560
CER(US$)	22.03	23.79	28.45	30.83	21.63	21.96	31.05	26.58	31.20	28.57	36.37	28.08
Decrease:5%												
Total Annual Cost (US$)	**1,120,312**	**1,163,916**	**1,412,500**	**1,474,736**	**1,016,857**	**1,001,241**	**1,392,205**	**1,195,376**	**1,346,258**	**1,241,541**	**1,585,705**	**1,251,276**
Change in Water Consumed per household/yr	58,400	58,400	58,400	58,400	52,560	52,560	52,560	52,560	52,560	52,560	52,560	52,560
CER(US$)	19.18	19.93	24.19	25.25	19.35	19.05	26.49	22.74	25.61	23.62	30.17	23.81
Increase: 15%												
Total Annual Cost (US$)	**1,506,656**	**1,614,708**	**1,910,300**	**2,126,168**	**1,311,241**	**1,307,433**	**1,871,965**	**1,598,488**	**1,933,441**	**1,761,333**	**2,137,465**	**1,702,188**
Change in Water Consumed per household/yr	58,400	58,400	58,400	58,400	52,560	52,560	52,560	52,560	52,560	52,560	52,560	52,560
CER(US$)	25.80	27.65	32.71	36.41	24.95	24.88	35.62	30.41	36.79	33.51	40.67	32.39
Sensitivity on Discount Rate & lifespan- Multi Way												
Base case: 15 years												
Total Annual	**1,286,484**	**1,389,312**	**1,661,400**	**1,800,352**	**1,137,049**	**1,154,337**	**1,632,085**	**1,396,832**	**1,639,849**	**1,501,437**	**1,911,585**	**1,476,132**

[8] It is necessary to recompute *crf in line with the variations in rates and lifespan years of the individual projects. The recomputed *crf has to be multiplied to project cost, and community mobilization cost to derive the new total annual cost in US$.

ost (US$)												
hange in Water onsumed per ousehold/yr	58,400	58,400	58,400	58,400	52,560	52,560	52,560	52,560	52,560	52,560	52,560	58,400
ER(US$)	22.03	23.79	28.45	30.83	21.63	21.96	31.05	26.58	31.20	28.57	36.37	28.08
ecrease:10 years												
otal Annual ost (US$)	**1,458,363**	**1,558,359**	**1,848,075**	**2,044,564**	**1,274,443**	**1,269,159**	**1,811,995**	**1,547,924**	**1,860,043**	**1,696,359**	**1,916,475**	**1,645,524**
hange in Water onsumed per ousehold/yr	58,400	58,400	58,400	58,400	52,560	52,560	52,560	52,560	52,560	52,560	52,560	52,560
ER(US$)	24.97	26.68	31.65	35.01	24.25	24.15	34.47	29.45	35.39	32.27	36.46	31.31
ncrease: 20 years												
otal Annual ost (US$)	**1,265,191**	**1,432,963**	**1,599,175**	**1,718,948**	**1,127,251**	**1,116,063**	**1,572,115**	**1,346,468**	**1,566,451**	**1,436,463**	**1,830,115**	**1,419,668**
hange in Water onsumed per ousehold/yr	58,400	58,400	58,400	58,400	52,560	52,560	52,560	52,560	52,560	52,560	52,560	52560
ER(US$)	21.66	24.54	27.38	29.43	21.45	21.23	29.91	25.62	29.80	27.33	34.82	27.01

Way and Multi-Way Sensitivity Analysis on CEA (Cont)
Source: Author's own construct, 2012.

Under the sensitivity analyses, the one-way varies only the discount rate while the multi-way varies both the discount rate and the lifespan of the interventions. Both analyses presents the worst case scenario reflected in the decrease in parameters and the best case scenario showing an increase in parameters.

Despite the fact that the variation of the discount rate in the one-way analysis shows some changes in the CER of the three interventions, these changes are not significant enough to deviate from earlier results in table 7. In both one-way and multi-way analyses, there was a close up between borehole and rain harvesting as both were seen to have had low CER. However, in terms of consistency, rain harvesting system recorded the lowest CER throughout the analyses. The results of the sensitivity tests/analyses reaffirms the stability of best and worst case scenarios.

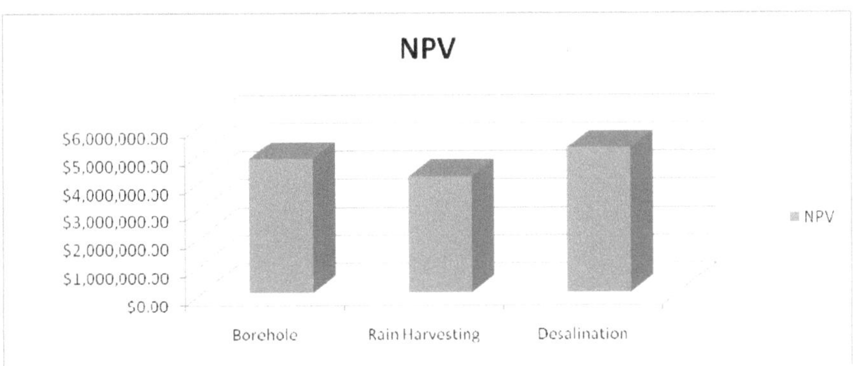

Figure 11: NPV comparison of three policy alternatives annualize costs
Source: Author's construct, 2012.

However, in terms of NPV, rain harvesting has the least Net Present Value stemming from the fact that, it had least total cost in comparison to the other two projects. Discounting all the three alternatives total costs to the present at a rate of 10% will still make rain harvesting as the best choice since cost to be expended is least. For instance, Borehole project had a total cost of $6,137,548.00 but with a NPV of $4,795,621.70, rain harvesting system had a total project cost of $5,320,303.00 with a NPV of $4,167,943.10. Desalination on the other hand had a total cost of $6,529,003.00 with a NPV of $5,176,048.96. The NPV scenarios also reaffirm earlier findings with regards to both the CER and sensitivity analyses[9] .

4.3 Conclusion

In Ghana, many rural communities and even urban communities do not have access to potable drinking water. Urban water supply is characterized by irregularities. Ghana therefore has major challenges in delivering potable water to its citizens. Water borne diseases has been some of the consequences of lack/inadequate supply of good drinking water. As a developing country, Ghana has to prioritize its development agenda to be able to address issues in all facet of the economy. Existing resources for each sector ought to be well harnessed for maximum benefits.

A cost effectiveness analysis is one such tool which can be used in making an informed decision to maximize results. This proposal sought to determine the most cost effective intervention between three policy alternatives; borehole, rain harvesting system and desalination plant in delivering potable water to households. These three interventions were considered because they are the current water supply interventions implemented in other developing countries. Ghana abounds in diverse water resources (rivers, streams, lakes, sea etc) that make it worth-while to consider these as policy alternatives both in the short and long runs. Though the analysis proved that rain water harvesting systems are the most cost effective than the rest given the same outcome, we therefore recommend this intervention to augment existing government efforts. With the continuous change in global climatic conditions and its attendant consequences, it will be suicidal on the part of any country to depend and pursue only one intervention as a source of water. Current technologies make it much easier to harness ocean water as a source of drinking water through the process of desalination. Coastal communities will benefit immensely from such intervention alongside those in landlocked areas. In doing this, government will be employing a multi interventional approach in tackling the challenges associated with water supply for both rural and urban communities. More effort is needed to address the perennial water shortages. As this policy proposal is aimed at achieving a 23.84% coverage from 2012 to 2015 (i.e. an average of 5.96% on yearly basis for 4 years), which is expected to push the national coverage rate from 52.16% to 76% by the close of the intervention period (2015).

4.4 Policy Recommendations

Based on the results of the cost effectiveness analysis, the following recommendations are essential for rural water supply in Ghana;

[9] See Appendix for NPV calculation.

- In the remote and sparsely populated areas where Government aims to supply water in the shortest possible time, rain harvesting systems should be provided as they are less expensive and do not require mammoth investments.
- Rain harvesting systems are maximized if the right population is served with the commensurate number of facilities. The right method in estimating the water needs of the beneficiary community should be well addressed to maximize the benefits of the system. In addition, threshold requirements for facilities provision should also be of concern.
- Diversification reduces risk tendencies and in line with the axiom that goes like "*don't put all your eggs in one basket*". The constant change in the rainfall pattern being a major threat to the provision of potable water raises the urgent need to seek for quick alternate way of getting access to safe drinking water, irrespective of its mammoth investment capital requirement. Having said this, we anticipate that the revenue from the oil industry which thrives at the coastal parts of the country should serve the purpose of providing desalination plants at the coast to serve the localities on whose lands the oil is being drilled. This could be in the form of social corporate responsibility or government stepping in to undertake such intervention. Lessons from Cyprus, Cape Verde and India have indicated how sustainable this intervention of desalination can be especially in the long term.

REFERENCES:

Alexander M., Alain R., Chris P., (2012). *Achieving financial sustainability and recovering costs in bank financed water supply and sanitation and irrigation projects*. Water papers www.worldbank/water

Anderson, J. E. (1994). *Public Policymaking: An Introduction* (2nd ed.). N.Y.: Houghton Mifflin

Ashlynn Stillwell (2008). *Desalination and Long-Haul Transfer as a Drinking Water Supply*.

Central Ground Water Board (2011). *Select case studies Rain water harvesting And Artificial recharge*. www.cgwb.gov.in

Dunn, William N. (2004). *Public Policy Analysis: An Introduction (3rd Ed.)*. N.J.: Prentice Hall.

Dye, Thomas R. (2008). *Understanding Public Policy (11th Ed.)*. N.J.: Prentice Hall.

Energy Recovery, Inc. (2011). *Minjur Seawater Desalination Plant*. www.energyrecovery.com

Faber Maunsell (2004). *Greywater Recycling & Rainwater Harvesting Feasibility Study Sustainable Eastside*.

Gtz/ Nkum Associates., (2003): "*PRA Guide on District Poverty Profiling and Mapping in Ghana*". A Hand Book.

Gyau-Boakye P. & Dapaah-Siakwan S., (1999). *Groundwater: solution to Ghana's rural water supply industry?* Water Resources Research Institute (CSIR), the Engineer. http://africantechnologyforum.com/GhIE/ruralwtr/ruralwtr.htm

Henry M., Levin & Patrick J. McEwan (2000): *Cost-effectiveness analysis as an evaluation tool.*

Jaime Sadhwani J., Jose Veza M., & Carmelo Santana (2005). *Case studies on environmental impact of seawater desalination*. Elservier press

Joana N.Guo (2011). *Cost effectiveness analysis of selected Water projects in Ghana, ISODEC*

Julio B., Martin-Ortega J., & Pascual M., (2010). *A Cost-Effectiveness Analysis of Water Saving Measures for the Water Framework Directive: the Case of the Guadalquivir River Basin in Southern Spain*. Springer Science Business Media B.V

Ministry of Water Resources, Works and Housing, CWSA (2008): *Strategic Investment Plan, 2008-2015 & the Medium Term Plan, 2008- 2012*. http://www.cwsagh.org/

Muhammad Akram Kahlown (2009). *Rainwater Harvesting in Cholistan Desert: A Case Study of Pakistan*. Pakistan Council of Research in Water Resources (PCRWR).

Norma Khoury-Nolde (2006). *Rainwater Harvesting*. Fbr Germany

Richard C. Carter (2009). *Towards cost-effective groundwater development in Sudan and in sub-Saharan Africa*. Personal Report of Workshop on Cost-effective Boreholes, Khartoum.

RWL Group (2008). *Case Study: Seawater Desalination*. www.nirosoft.com

The Schumacher Centre of Technology& Development (2008). *Rainwater harvesting*. www.practicalaction.org

Thomas Clasen F. & Laurence Haller (2008). *Water Quality Interventions to Prevent Diarrhoea: Cost and Cost-Effectiveness*. WHO press. http://water.org/water-crisis/water-facts/water

Weimer, David L. and Vining, Aidan R. (2011). *Policy Analysis: Concepts and Practice* (5th Ed.). New Jersey: Prentice Hall.

Water and Sanitation Program (2006). *Ten-step Guide towards Cost-effective Boreholes Case study of drilling costs in Ethiopia.*

WHO (2009). *Diarrhoea: why children are still dying and what can be done.* WHO & UNICEF Press. http://apps.who.int/iris/bitstream/10665/44174/1/9789241598415_eng.pdf

APPENDICES:

Appendix 1: Calculation of Water Consumed **Before** and **After** Interventions **(Rain harvesting system and Desalination plant).**

2012	Population= ***1,564,875***, household size=8 Improved water consumption per capita per day before intervention= 10 litres Therefore water consumed by population per year=1,564,875*10*365=5,711,793,750 No. of Households= 1,564,875/8 = 195609.375 Water consumed per household per year ***before*** intervention= 5,711,793,750/195609.375= 29,200 Water consumed per capita per day after intervention = 20 litres for Rain harvesting stand pipes and 60 litres for home connection. NOTE: Project design allows for 20% home connection and 80% standpipes. 20% of population= 0.2*1,564,875= 312,975 80% of population = 1,251,900 312,975*60*365=6,854,152,500 1,251,900*20*365=9,138,870,000 Water consumed after intervention= 6,854,152,500+9,138,870,000= 15993022500 Improved water consumed per household per year ***after*** intervention= 15993022500/195609.375 = 81,760 **Change in improved water consumed= 81,760-29,200= 52,560**
2013	Population= ***1,471,608***, household size=8 Improved water consumption per capita per day before intervention= 10 litres Therefore water consumed by population per year=1,471,608*10*365=5,371,369,200 No. of Households= 1,471,608/8 = 183,951 Water consumed per household per year ***before*** intervention= 5,371,369,200/183,951= 29,200 Water consumed per capita per day after intervention = 20 litres for Rain harvesting stand pipes and 60 litres for home connection.

	NOTE: Project design allows for 20% home connection and 80% standpipes. 20% of population= 0.2*1,471,608= 294,321.60 80% of population = 1,177,286.40 294,321.60*60*365=6,445,643,040 1,177,286.40*20*365=8,594,190,720 Water consumed after intervention= 6,445,643,040+8,594,190,720= 15,039,833,760 Improved water consumed per household per year ***after*** intervention= 15,039,833,760/183,951 = 81,760 **Change in improved water consumed= 81,760-29,200= 52,560**
2014	Population= ***1,383,900***, household size=8 Improved water consumption per capita per day before intervention= 10 litres Therefore water consumed by population per year=1,383,900*10*365=5,051,235,000 No. of Households= 1,383,900/8 = 172,987.50 Water consumed per household per year ***before*** intervention= 5,051,235/172,987.50= 29,200 Water consumed per capita per day after intervention = 20 litres for Rain harvesting stand pipes and 60 litres for home connection. ***NOTE***: Project design allows for 20% home connection and 80% standpipes. 20% of population= 0.2*1,383,900= 276,780.00 80% of population = 1,107,120 276,780*60*365=6,061,482,000 1,107,120*20*365=8,081,976,000 Water consumed after intervention= 6,061,482,000+8,081,976,000= 14,143,458,000 Improved water consumed per household per year ***after*** intervention= 14,143,458,000/172,987.50 = 81,760 **Change in improved water consumed= 81,760-29,200= 52,560**

2015	Population= ***1,301,420***, household size=8 Improved water consumption per capita per day before intervention= 10 litres Therefore water consumed by population per year=1,301,420*10*365=4,750,183,000 No. of Households= 1,301,420/8 = 162,677.50 Water consumed per household per year ***before*** intervention= 4,750,183,000/162,677.50= 29,200 Water consumed per capita per day after intervention = 20 litres for Rain harvesting stand pipes and 60 litres for home connection. ***NOTE***: Project design allows for 20% home connection and 80% standpipes. 20% of population= 0.2*1,301,420= 260,284 80% of population = 1,041,136 260,284*60*365=5,700,219,600 1,041,136*20*365=7,600,292,800 Water consumed after intervention= 5,700,219,600+7,600,292,800= 13,300,512,400 Improved water consumed per household per year ***after*** intervention= 13,300,512,400/162,677.50 = 81,760 **Change in improved water consumed= 81,760-29,200= 52,560**

Source: Author's construct, 2012.

Appendix 2: Calculation of Water Consumed Before and After Interventions **(Borehole)**

2012	Population= ***1,564,875***, household size=8 Number of households= 1,564,875/8= 195,609.38 Improved water consumption per capita per day before intervention= 0 Improved water consumption per capita per day after intervention= 20 litres Water consumed by population per year = 1,564,875*20*365=11,423,587,500 Water consumed per household per year = 11,423,587,500/195,609.38= 58399.99851 **Change in improved water consumed= 58399.99851-0= 58,400**

2013	Population= ***1,471,608***, household size=8 Number of households= 1,471,608/8= 183,951 Improved water consumption per capita per day before intervention= 0 Improved water consumption per capita per day after intervention= 20 litres Water consumed by population per year = 1,471,608*20*365=10,742,738,400 Water consumed per household per year = 10,742,738,400/183,951=58,400 **Change in improved water consumed= 58,400-0= 58,400**
2014	Population= ***1,383,900***, household size=8 Number of households= 1,383,900/8= 172,987.50 Improved water consumption per capita per day before intervention= 0 Improved water consumption per capita per day after intervention= 20 litres Water consumed by population per year = 1,383,900*20*365=10,102,470,000 Water consumed per household per year = 10,102,470,000/172,987.50=58,400 **Change in improved water consumed= 58,400-0= 58,400**
2015	Population= ***1,301,420***, household size=8 Number of households= 1,301,420/8= 162,677.50 Improved water consumption per capita per day before intervention= 0 Improved water consumption per capita per day after intervention= 20 litres Water consumed by population per year = 1,301,420*20*365=9,500,366,000 Water consumed per household per year = 9,500,366,000/162,677.50=58,400 **Change in improved water consumed= 58,400-0= 58,400**

Source: Author's construct, 2012.

Appendix 3: NPV calculation

Calculation of Net Present Value of cost of **Borehole project** over 4 years at 10% Discount rate (in million dollars)

Year	Future Value (FV)	Discount Rate (r)	Number of Years (n)	Discount Factor (DF) $[1/(1+r)^n$	Present value
2012	$1,286,484.00	0.10	1	$0.91	$1,169,530.91
2013	$1,389,312.00	0.10	2	$0.83	$1,148,191.74
2014	$1,661,400.00	0.10	3	$0.75	$1,248,234.41
2015	$1,800,352.00	0.10	4	$0.68	$1,229,664.64
	$6,137,548.00				**$4,795,621.70**

Source: Author's construct, 2012.

Calculation of Net Present Value of annualize cost of **Rain harvesting project** over 4 years at 10% Discount rate (in million dollars)

Year	Future Value (FV)	Discount Rate (r)	Number of Years (n)	Discount Factor (DF) $[1/(1+r)^n$	Present value
2012	$1,137,049.00	0.10	1	$0.91	$1,033,680.91
2013	$1,154,337.00	0.10	2	$0.83	$953,997.52
2014	$1,632,085.00	0.10	3	$0.75	$1,226,209.62
2015	$1,396,832.00	0.10	4	$0.68	$954,055.05
	$5,320,303.00				**$4,167,943.10**

Source: Author's construct, 2012.

Calculation of Net Present Value of annualize cost of **Desalination project** over 4 years at 10% Discount rate (in million dollars)

Year	Future Value (FV)	Discount Rate (r)	Number of Years (n)	Discount Factor (DF) $[1/(1+r)^n$	Present value
2012	$1,639,849.00	0.10	1	$0.91	$1,490,771.82
2013	$1,501,437.00	0.10	2	$0.83	$1,240,857.02

2014	$1,911,585.00	0.10	3	$0.75	$1,436,202.10
2015	$1,476,132.00	0.10	4	$0.68	$1,008,218.02
	$6,529,003.00				**$5,176,048.96**

Source: Author's construct, 2012.